Galaxies

Lily Erlic

DEEP in SPACE

www.av2books.com

Go to www.av2books.com, and enter this book's unique code.

BOOK CODE

AVS38732

AV² by Weigl brings you media enhanced books that support active learning.

AV² provides enriched content that supplements and complements this book. Weigl's AV² books strive to create inspired learning and engage young minds in a total learning experience.

Your AV² Media Enhanced books come alive with...

Audio
Listen to sections of the book read aloud.

Video
Watch informative video clips.

Embedded Weblinks
Gain additional information for research.

Try This!
Complete activities and hands-on experiments.

Key Words
Study vocabulary, and complete a matching word activity.

Quizzes
Test your knowledge.

Slideshow
View images and captions, and prepare a presentation.

... and much, much more!

Published by AV² by Weigl
350 5th Avenue, 59th Floor
New York, NY 10118
Website: www.av2books.com

Library of Congress Control Number: 2019941860

ISBN 978-1-7911-0970-7 (hardcover)
ISBN 978-1-7911-0971-4 (softcover)
ISBN 978-1-7911-0972-1 (multi-user eBook)

Printed in Guangzhou, China
1 2 3 4 5 6 7 8 9 0 23 22 21 20 19

052019
102918

Project Coordinator: John Willis
Designer: Terry Paulhus

Weigl acknowledges Alamy, Getty Images, and Wikimedia as its primary image suppliers for this title.

CONTENTS

4

Galaxies are large groups of stars. They also have gas, dust, and planets. Earth is in a galaxy called the Milky Way.

There are three kinds of galaxies. They are spiral, elliptical, and irregular.

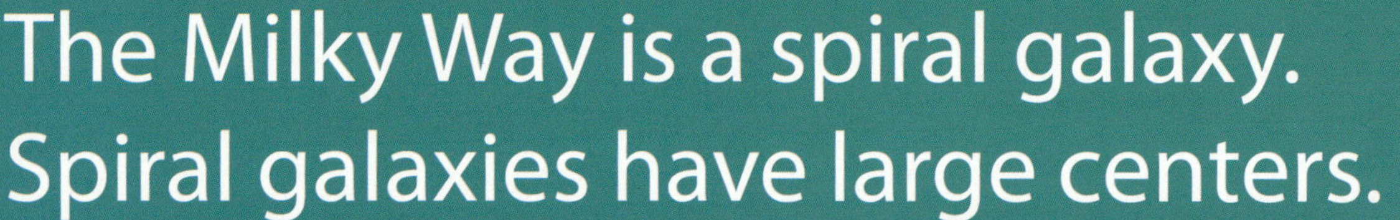

The Milky Way is a spiral galaxy.
Spiral galaxies have large centers.

They have long arms made of stars and gas. The arms move around the center of each galaxy.

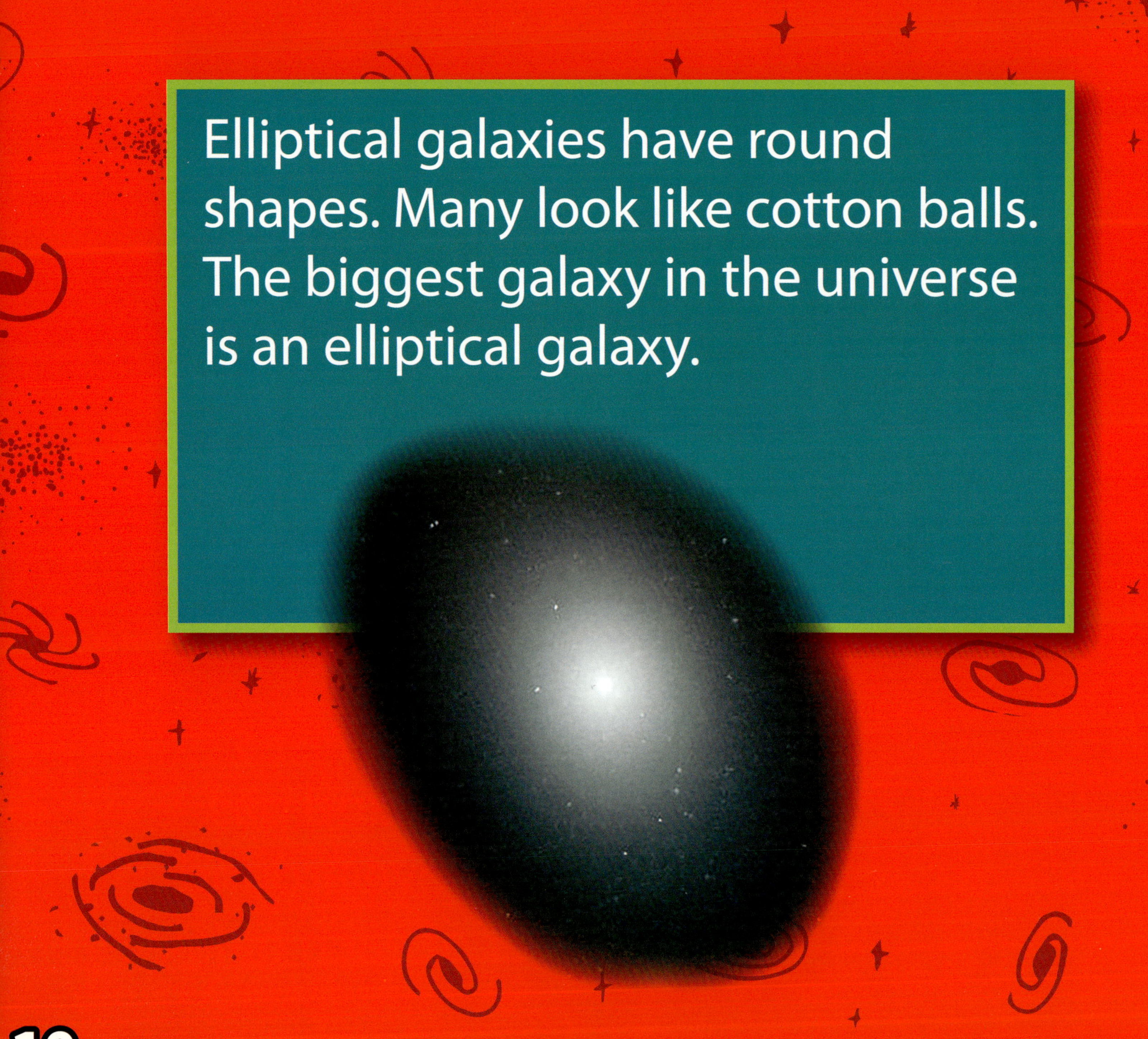

Elliptical galaxies have round shapes. Many look like cotton balls. The biggest galaxy in the universe is an elliptical galaxy.

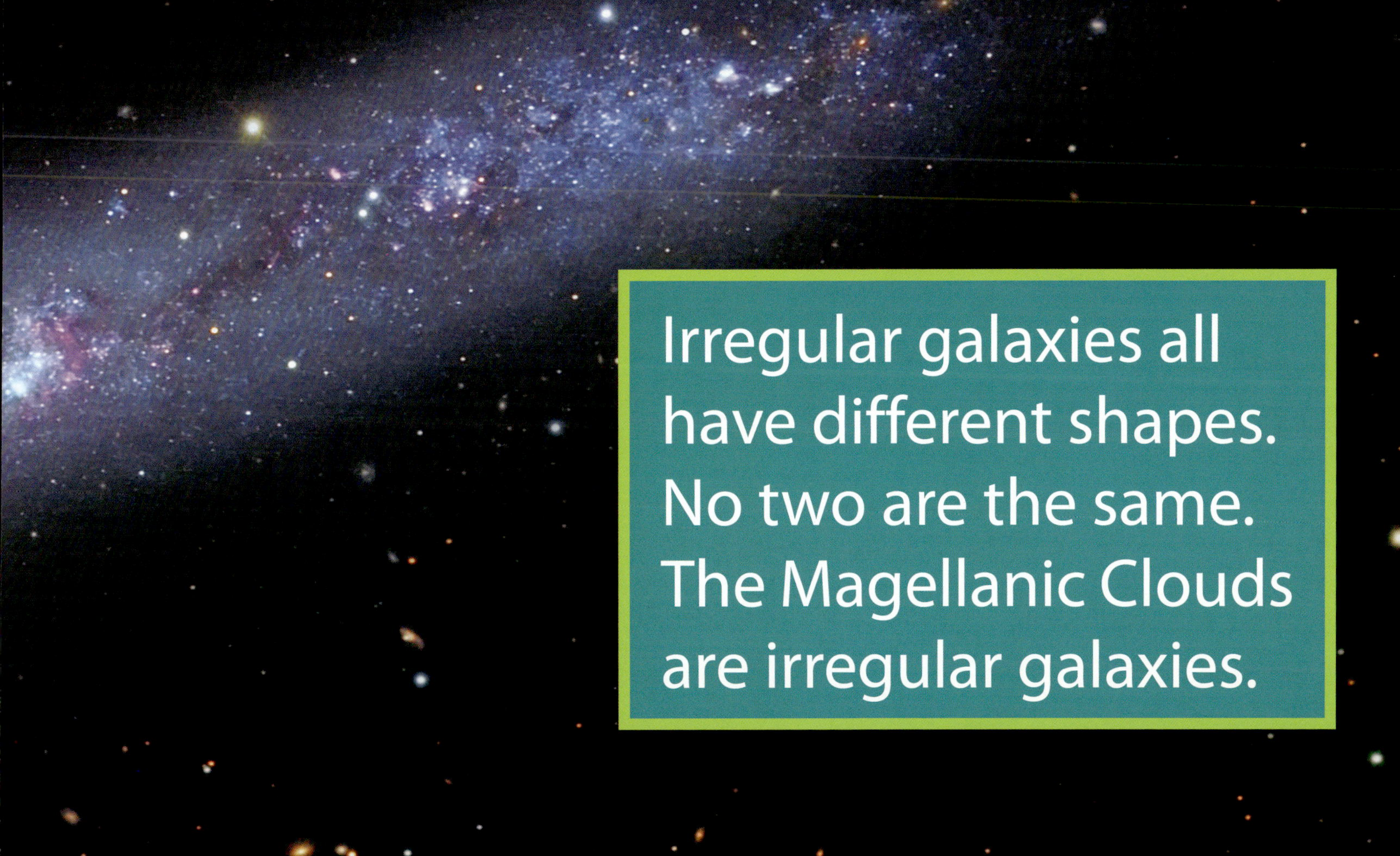

Irregular galaxies all have different shapes. No two are the same. The Magellanic Clouds are irregular galaxies.

Scientists think most galaxies have a large black hole in their center. Stars in a galaxy spin around the black hole.

There may be more than 100 billion galaxies in the universe.
The Coma Cluster is a group of thousands of them.

Every galaxy is different. Pictures of galaxies often look like the insides of kaleidoscopes.

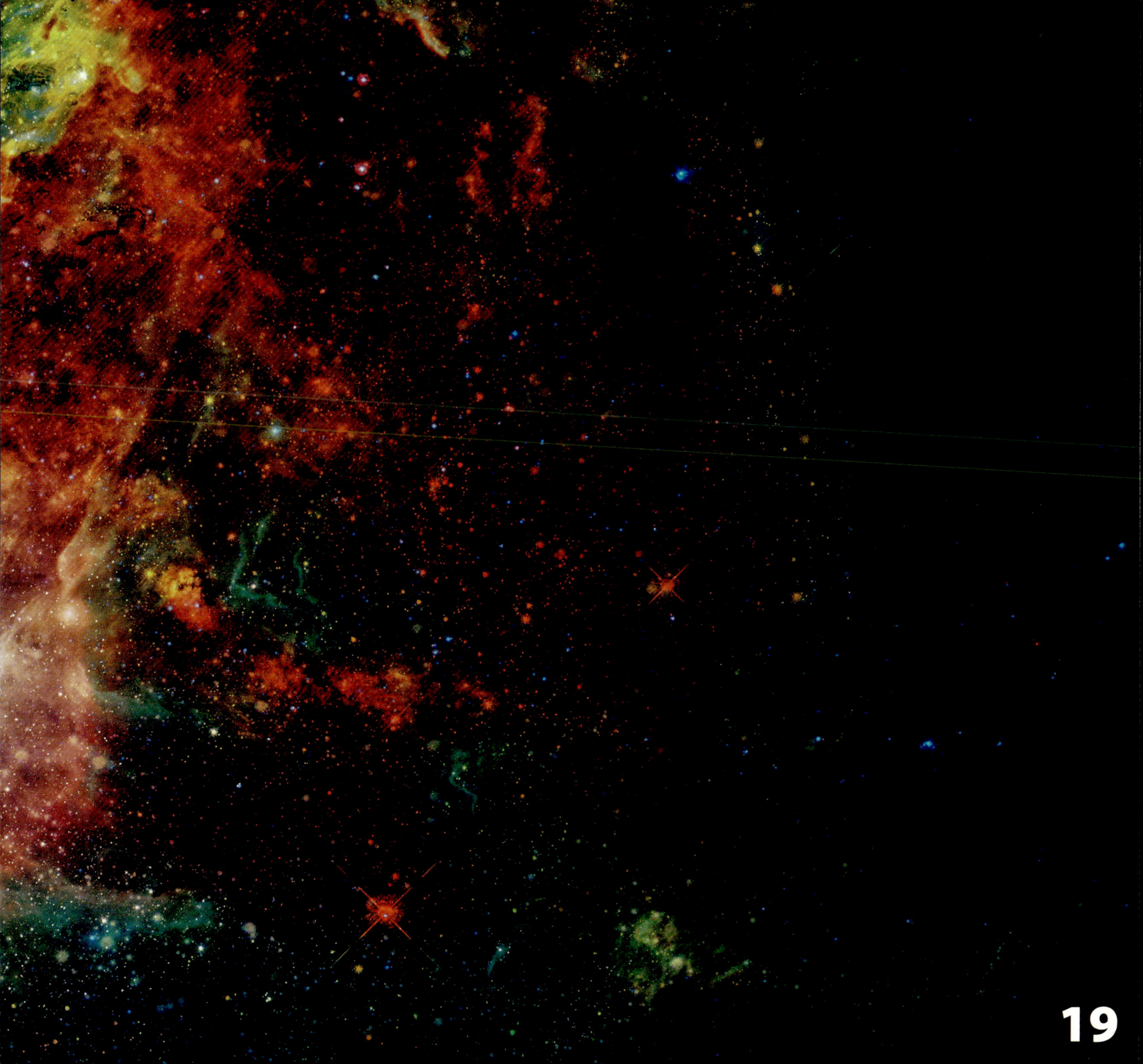

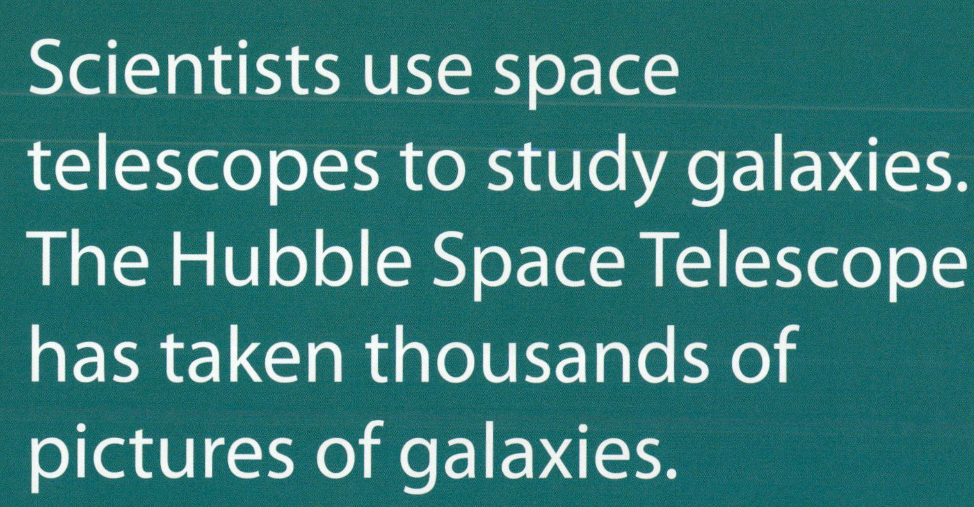

Scientists use space telescopes to study galaxies. The Hubble Space Telescope has taken thousands of pictures of galaxies.

GALAXY FACTS

These pages provide detailed information that expands on the interesting facts found in the book. They are intended to be used by adults as a learning support to help young readers round out their knowledge of each object or event featured in the *Deep in Space* series.

Pages 4–5

Galaxies are large groups of stars. The word "galaxy" comes from the Greek word *galaxias*, which means "milky." Because there are so many stars in galaxies, they look like pools of spilled milk in space. Every object in space has gravity. This force holds the stars in a galaxy together. Gravity is on Earth, too. It holds people and objects on the ground.

Pages 6–7

There are three kinds of galaxies. A scientist named James Hubble identified the types of galaxies. He created a list of characteristics for each one. When scientists discover a new galaxy, they use this list to decide what kind it is. Spiral and elliptical galaxies are the most common types.

Pages 8–9

The Milky Way is a spiral galaxy. A spiral galaxy can be ordinary or barred. An ordinary spiral galaxy has a round center. The center of a barred spiral galaxy is long and flat-looking. The spiral arms give these galaxies their shape. The Milky Way is a barred spiral galaxy. It is at least 100,000 light years across. One light year is equal to almost 6 trillion miles (10 trillion kilometers).

Pages 10–11

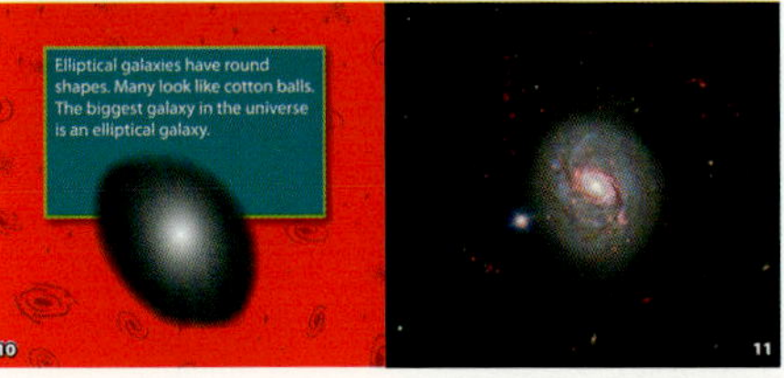

Elliptical galaxies have round shapes. Some elliptical galaxies look like a perfect circle. Others look like eggs. Some even look like logs with rounded ends. Elliptical galaxies do not have much gas or dust in them. They contain older stars. Unlike spiral galaxies, elliptical galaxies do not have a bright center. The biggest known elliptical galaxy is called IC 1101.

Pages 12–13

Irregular galaxies all have different shapes. About one out of every four galaxies is irregular. Some irregular galaxies do not have any shape or structure at all. Others have unique shapes. The Magellanic Clouds fall into this category. The Small Magellanic Cloud is about 210,000 light years away from the Milky Way. The Large Magellanic Cloud is one of the nearest galaxies to the Milky Way. It is 179,000 light years away.

Pages 14–15

Scientists think most galaxies have a large black hole in their center. A black hole is an area in space that pulls stars and other objects inside of it. The force of gravity in a black hole is very strong. Scientists know that black holes spin very fast because they can measure how fast stars and other objects spin around them. Although scientists cannot see black holes, they can see stars and gas clouds stretching into them.

Pages 16–17

There may be more than 100 billion galaxies in the universe. Scientists count galaxies in an area of space. They use the number to predict how many more galaxies there could be. The Coma Cluster is one group of galaxies. The galaxies in the cluster are very close together. The Coma Cluster is about 300 million light years from Earth.

Pages 18–19

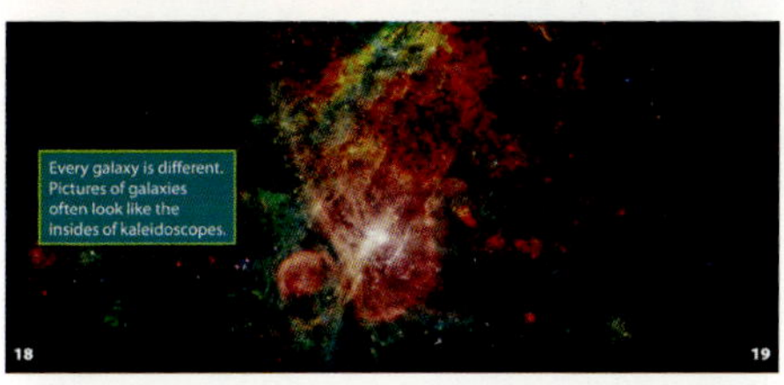

Every galaxy is different. Scientists use images from space to show people what galaxies look like. Telescopes take many photos of a galaxy. Then, the pictures are put together to make one picture, called a composite. The photos are black and white. Scientists add colors to the photos. The colors stand for different things, such as kinds of gases, or light and darkness. The colors make these photos look like kaleidoscope designs.

Pages 20–21

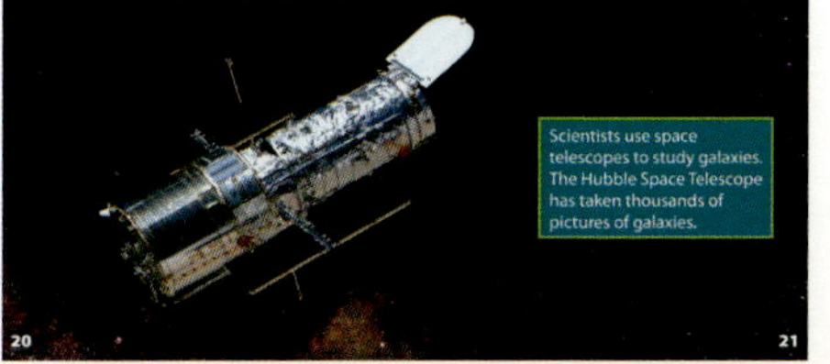

Scientists use space telescopes to study galaxies. The National Aeronautics and Space Administration (NASA) has several telescopes in space. Some orbit the Sun and some orbit Earth. Most pictures of galaxies are taken by space telescopes. The Hubble Space Telescope has taken more than 1 million photos of space objects and galaxies. It has allowed humans to see distant parts of space.

KEY WORDS

Research has shown that as much as 65 percent of all written material published in English is made up of 300 words. These 300 words cannot be taught using pictures or learned by sounding them out. They must be recognized by sight. This book contains 45 common sight words to help young readers improve their reading fluency and comprehension. This book also teaches young readers several important content words, such as proper nouns. These words are paired with pictures to aid in learning and improve understanding.

Page	Sight Words First Appearance
5	a, also, and, are, Earth, groups, have in, is, large, of, the, they
6	kinds, there, three
9	around, each, long, made, move
10	an, like, look, many
13	all, different, no, two, same
14	most, their, think
16	be, may, more, than, them
18	every, often, pictures
21	has, study, to, use

Page	Content Words First Appearance
5	dust, galaxies, gas, Milky Way, planets, stars
6	elliptical, irregular, spiral
8	centers
9	arms
10	cotton balls, universe
13	Magellanic Clouds
14	black hole, scientists
16	Coma Cluster
18	kaleidoscopes
21	Hubble Space Telescope